Understanding the Elements of the Periodic Table™

MANGANESE

Heather Hasan

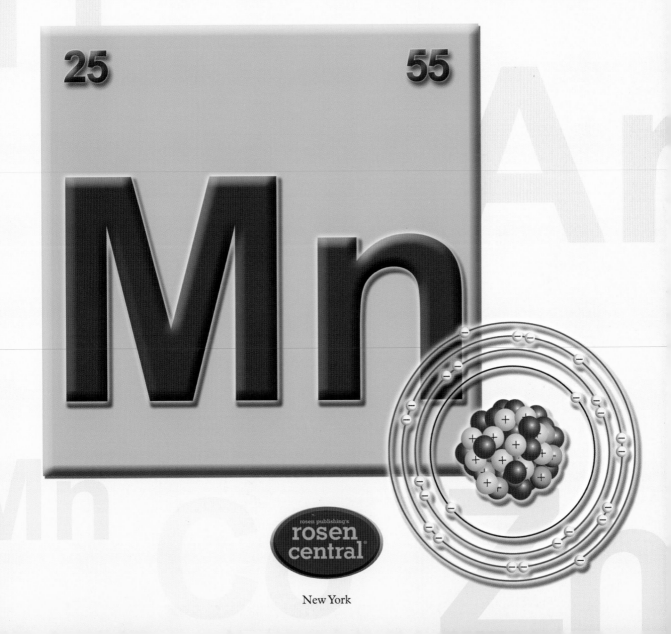

25

55

Mn

rosen publishing's
rosen central®

New York

For my family: Omar, Samuel, and Matthew,
you bring me indescribable joy

Published in 2008 by The Rosen Publishing Group, Inc.
29 East 21st Street, New York, NY 10010

Library of Congress Cataloging-in-Publication Data

Hasan, Heather.
Manganese / Heather Hasan.—1st ed.
    p. cm.—(Understanding the elements of the periodic table)
Includes bibliographical references and index.
ISBN-13: 978-1-4042-1408-8 (library binding)
1. Manganese. 2. Periodic law. 3. Chemical elements. I. Title.
QD181.M6H37 2008
546'.541—dc22

                                                                2007027121

Manufactured in Malaysia

On the cover: Manganese's square on the periodic table of elements; *(inset)* model of the subatomic structure of a manganese atom.

Contents

Introduction

In 480 BCE, the Persian Empire was preparing to conquer all of Greece. While many of their countrymen fled, a band of Spartan warriors courageously fought against the largest army the world had ever seen. Fighting fiercely, the Spartans were able to hold the Persian army back for three days at Thermopylae, a narrow valley pass near present-day Lamia.

In ancient Greece, all other armies paled in comparison to the Spartans. Fittingly, the name "Spartan" means warlike. In their militaristic culture, Spartan men were raised from birth to be warriors. Weak or small infants were abandoned to die. At the age of seven, Spartan boys were sent from home to live and train with other young boys. Only after proving themselves brave and crafty were they allowed to take up the Spartan sword and shield and join the ranks of one of the most formidable armies ever.

The shields that the Spartan soldiers carried were so large that they were also used as stretchers to remove the wounded from the battlefield. The Spartans protected their heads with large metal helmets, and even their legs were covered by metal leg guards. The Spartans carried long swords in their hands and wore short swords on their waists for close combat.

The metal the Spartans used for their armor, swords, and shields was unusually hard. Analysis shows that manganese (chemical symbol: Mn) is found in the iron ores, or minerals, that the Spartans used to make their

This illustration shows the Spartan army in battle. The inset shows a Spartan foot soldier with metal shield, spear, helmet, breastplate, and leg armor. Manganese in this metal made it very strong.

metal. Because of this, historians think the Spartans unknowingly made their weapons out of a mixture of iron (Fe) and manganese metal. Today, we know that this combination makes a very strong metal. With the help of manganese, the Spartans wielded exceptionally strong weapons—fitting for such brave warriors!

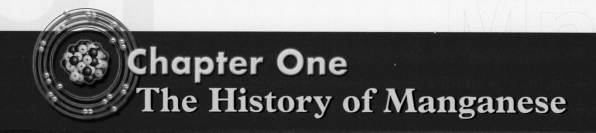

Chapter One
The History of Manganese

Manganese was recognized as an element in the late 1700s. However, people were using manganese compounds (manganese combined with other elements) long before that. In fact, we know that manganese-containing minerals have been used since prehistoric times. Cave paintings made with such minerals can be traced back 17,000 years. Several thousand years ago, ancient Egyptian artisans used manganese compounds, too, in the shiny protective glaze they put on their pottery. Glassmakers have also used manganese compounds since early times. Both the Egyptians and Romans used manganese compounds to either remove color from glass or to add color to it. To this day, manganese compounds are used to color glass.

Where Manganese Is Found

Manganese is found everywhere. Traces of it are found in soil, water, plants, and animals. It is also found in meteorites that have fallen to the earth from outer space. Manganese is not found in its pure state except in these meteorites. In the soil, manganese is found in many minerals, and it is the twelfth most common element found in the earth's crust. Each year, about 20 million tons of manganese ore is mined around the world. The

Paintings in the Altamira caves in Spain show horses, wild boars, stags, and bison (one of which is shown here). The paintings were made by prehistoric people using manganese and other minerals.

element is also found at the bottom of the sea. Deep-sea exploration has revealed that patches of the ocean floor are covered with manganese nodules, or lumps, about the size of potatoes.

Scientists think most of the manganese ore mined today was formed deep underwater. They believe that hot volcanic rock deep in the ocean heated the water around it. This warm water caused manganese in the rock to dissolve into the water. The manganese then floated around until ocean currents brought it to shallow water, which is rich in oxygen. The manganese combined with the oxygen and formed solid minerals. These minerals settled on the seabed and over a long time hardened into rock.

This manganese deposit is being mined in Burkina Faso, a country located in West Africa. The economy of Burkina Faso relies on developing the country's few mineral resources.

The Naming of Manganese

The name "manganese" comes from *magnes*, the Latin word for magnet. In ancient times, two black minerals from Magnesia (in present-day Greece) were both called magnes. One was considered to be the male magnes and was magnetic. This mineral is now known as magnetite, an ore from which we get iron. The other black mineral was considered the female magnes. It was not magnetic. This feminine magnes, later called magnesia, is now known as pyrolusite or manganese dioxide. It is one of the ores from which we get manganese.

The Discovery of Manganese

Many elements, such as manganese, have been used since ancient times in their compound forms. But it was not until the late eighteenth century that the modern idea of individual chemical elements really took hold. Around that time, scientists began to discover that well-known minerals were made up of several elements. In 1774, Swedish chemist Carl Wilhelm Scheele (1742–1786) was working with the mineral pyrolusite (MnO_2). Following some experiments, he suggested that manganese might be a chemical element. In addition to manganese, Scheele suggested that oxygen (O), chlorine (Cl), and barium (Ba) were elements, too. That is four new elements! Though Scheele himself was not able to isolate manganese, his work went a long way toward identifying it as an element.

It was up to another scientist to produce the first pure sample of the element manganese. Johan Gottlieb Gahn (1745–1818), a Swedish chemist and Scheele's friend, finally isolated the element in 1774. He obtained it by heating the mineral pyrolusite in the presence of charcoal. In this way, Gahn produced a small button of grayish manganese. Manganese was officially an element!

Johan Gottlieb Gahn began his career as a miner before he began studying mineralogy. Gahnite, a rare, deep-blue mineral, is named after him.

The History of the Periodic Table

Today, we know of more than 100 elements. As more and more elements were discovered, scientists needed to organize them. One of these scientists was Dmitry Mendeleyev (1834–1907), a Russian chemist teaching at the

War Nickels

Today's nickels are made of a mixture of 25 percent nickel (Ni) and 75 percent copper (Cu). During World War II, however, copper and nickel were needed for wartime purposes. So, between the years 1942 and 1945, U.S. nickels were made with a mixture of 56 percent copper, 35 percent silver (Ag), and 9 percent manganese. See if you can find one of these special "war nickels." They were the only U.S. coins made with manganese until 2000, when the Sacagawea dollar was released. This golden-colored coin is made from 88 percent copper, 6 percent zinc (Zn), 2 percent nickel, and 4 percent manganese. The image on the coin depicts Sacagawea, the Shoshone Indian woman who assisted Lewis and Clark on their famous expedition.

University of St. Petersburg. In 1869, he began organizing the elements in a way that would make them easier for his students to learn and understand. He listed the elements that were known at that time in order of their atomic weights. Upon arranging the elements in this way, Mendeleyev noticed that certain sets of properties, or characteristics, of the elements created a pattern in the list. He then made a chart of the elements in order of atomic weights, with the elements that have similar properties arranged in columns. Because the properties of the elements on the table vary in periods, or recurring patterns, the table that Mendeleyev created became known as the periodic table.

Manganese was one of the elements listed on Mendeleyev's original table. However, Mendeleyev did not know all of the elements that we know of today. The Russian chemist wisely left gaps where he believed undiscovered elements belonged. And based on the locations of those gaps in the table, he was able to predict with great accuracy the properties of missing elements that were later discovered.

Chapter Two
The Element Manganese

In scientific terms, something that has mass and takes up space is called matter. All the matter in the universe—including the air you breathe, the ground you walk on, and even you—is made up of atoms. An element, such as manganese, is a special substance that is made up of only one kind of atom. All atoms in pure manganese are exactly the same. Atoms are very tiny. They are so tiny that it would take about 40 million manganese atoms, lying side by side, to form a line that is only 0.4 inches (1 centimeter) long! Amazingly, atoms are made up of even smaller parts called subatomic particles. In order to understand what makes manganese unique, one must take a closer look at these particles.

Subatomic Particles

There are three basic building blocks in an atom: protons, neutrons, and electrons. The number and arrangement of these subatomic particles determine the properties of an element. Protons and neutrons are clustered together, forming the nucleus at the center of an atom. As protons and neutrons are much heavier than electrons, the nucleus contains almost all the atom's mass. Neutrons have no electrical charge, while the protons have a positive electrical charge. This gives the nucleus an overall positive electrical charge. Since the nucleus of a manganese atom has twenty-five

This model of a manganese atom shows twenty-five protons and thirty neutrons clustered together in the nucleus. Twenty-five electrons are positioned in four shells surrounding the atom's nucleus.

protons, it has a nuclear charge of +25. Atoms of manganese are unique because no other element has this same number of protons.

Electrons are negatively charged particles. They are arranged in layers, or shells, around the positively charged nucleus of an atom. Electrons are not fixed in position but move rapidly around the nucleus. The number of electrons surrounding the nucleus is equal to the number of protons in the nucleus. This balances the electrical charge of the atom, making the net charge of the atom zero. Such an atom is said to be neutral. A neutral manganese atom with twenty-five protons in its nucleus also has twenty-five electrons circulating in the shells surrounding its nucleus.

Isotopes and Atomic Weight

The number of protons in an atom of a particular element is always the same—that's what makes it unique. However, in many elements, atoms exist in different forms called isotopes. Isotopes have the same number of protons and electrons but have different numbers of neutrons. In nature, almost all of the elements exist as a mixture of two or more isotopes. Manganese, however, has only one naturally occurring isotope, which has thirty neutrons. Several other manganese isotopes do exist, but they are not important to our discussion here.

Dmitry Mendeleyev arranged the elements in his table according to atomic weight. This is the average weight of all the isotopes of an element, taking into consideration how often each isotope occurs naturally. Because it is an average number, the atomic weight of most elements is a decimal number. Manganese, for example, has an atomic weight of about 54.938 atomic mass units (amu). On periodic tables, this more precise number is often rounded to 55 for convenience.

Manganese and the Periodic Table

Unlike Mendeleyev's periodic table, the one we use today lists the elements in order of increasing atomic number. This is the number of protons in the nucleus. The periodic table is an important tool for scientists. Among other things, it allows them to predict the properties of an element based on its location on the table. Take the metals, for example. Except for hydrogen (H), all the elements located on the left of the periodic table are metals. In the periodic table in the back of this book (see pages 40–41), metals are shown in orange, red, green, and blue-gray. Metals are easily recognized by their physical traits. Generally, metals can be polished to be made shiny. They also conduct electricity. Most metals have the ability to be hammered into shapes without breaking. This is called malleability. Metals are also usually ductile, meaning that they are able to be pulled into wires.

The elements located on the right in the table are called nonmetals. On the periodic table in this book, the nonmetals are shown in yellow and light gray. Elements such as carbon (C), oxygen, and chlorine are nonmetals because they lack the characteristics of metals. Between the metals and the nonmetals are the metalloids. Metalloids, or semimetals, have some characteristics similar to metals and others that are more like nonmetals. The metalloids are boron (B), silicon (Si), germanium (Ge), arsenic (As), antimony (Sb), tellurium (Te), and polonium (Po).

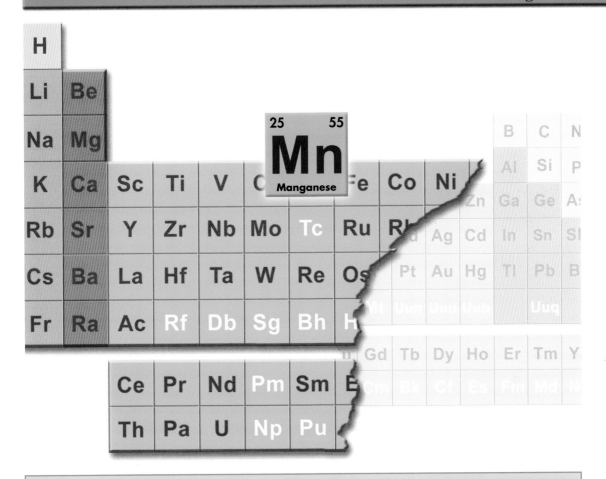

Manganese is a transition metal found on the left of the periodic table. Manganese's square in the periodic table often lists the element's atomic number (25) and its approximate atomic weight (55), along with its chemical symbol (Mn).

Groups and Periods

In the periodic table, each horizontal row of elements is called a period. The elements are arranged in order of increasing atomic number, so the number increases by one from one element to the next. All of the elements in a period have the same number of electron shells surrounding the nucleus of their atoms. Manganese is in period 4, indicating that it has four shells of electrons surrounding its nucleus. (See illustration on page 13.)

← Group →

	IA 1	IIA 2	IIIB 3	IVB 4	VB 5	VIB 6	VIIB 7	VIIIB 8	VIIIB 9	
1	1 **H** 1 Hydrogen									
2	3 **Li** 7 Lithium	4 **Be** 9 Beryllium								
3	11 **Na** 23 Sodium	12 **Mg** 24 Magnesium								
4	19 **K** 39 Potassium	20 **Ca** 40 Calcium	21 **Sc** 45 Scandium	22 **Ti** 48 Titanium	23 **V** 51 Vanadium	24 **Cr** 52 Chromium	25 **Mn** 55 Manganese	26 **Fe** 56 Iron	27 **Co** 59 Cobalt	Ni
5	37 **Rb** 85 Rubidium	38 **Sr** 88 Strontium	39 **Y** 89 Yttrium	40 **Zr** 91 Zirconium	41 **Nb** 93 Niobium	42 **Mo** 96 Molybdenum	43 **Tc** 98 Technetium	44 **Ru** 101 Ruthenium	45 **Rh** 103 Rhodium	Pd
6	55 **Cs** 133 Cesium	56 **Ba** 137 Barium	57 **La** 139 Lanthanum	72 **Hf** 178 Hafnium	73 **Ta** 181 Tantalum	74 **W** 184 Tungsten	75 **Re** 186 Rhenium	76 **Os** 190 Osmium	77 **Ir** 192 Iridium	Pt
7	87 **Fr** 223 Francium	88 **Ra** 226 Radium	89 **Ac** 227 Actinium	104 **Rf** 261 Rutherfordium	105 **Db** 262 Dubnium	106 **Sg** 266 Seaborgium	107 **Bh** 264 Bohrium	108 **Hs** 277 Hassium	109 **Mt** 268 Meitnerium	

Period

58 **Ce** 140 Cerium	59 **Pr** 141 Praseodymium	60 **Nd** 144 Neodymium	61 **Pm** 145 Promethium	62 **Sm** 150 Samarium	63 **Eu** 152 Europium	64 **Gd** 157 Gadolinium	Tb
90 **Th** 232 Thorium	91 **Pa** 231 Protactinium	92 **U** 238 Uranium	93 **Np** 237 Neptunium	94 **Pu** 244 Plutonium	95 **Am** 243 Americium	96 **Cm** 247 Curium	

Manganese is in period 4 on the periodic table. The element's period indicates that it has four electron shells surrounding its nucleus. Manganese is also in group VIIB (or group 7). All of the elements in group VIIB have two electrons in their outermost electron shells.

There are two electrons in manganese's innermost electron shell, eight in the second shell, thirteen in the third shell, and two in its outermost shell. The electrons in the outmost shell are called valence electrons. These valence electrons determine how an element like manganese behaves.

Each vertical column of elements in the table is called a group. As you go down a group, the elements get heavier. All of the elements in a group have the same number of electrons in their outermost electron shell. Because of this, the elements in a given group react in similar ways. Manganese is found in group VIIB (or group 7), along with technetium (Tc), rhenium (Re), and bohrium (Bh).

Manganese and the Transition Metals

Manganese and the other group VIIB elements are sometimes called the manganese family of elements. They are classified along with a larger group of elements called the transition metals. Transition metals are located in the middle of the periodic table, between groups IIA (also called group 2) and IIIA (group 13). They serve as a bridge, or transition, between the two sides of the table. They are shown in green in the periodic table in the back of this book. Transition metals are different from the elements in A groups because they have empty spaces not only in their outermost electron shell but also in the second from the outermost shell. Because of this electron arrangement, transition metals share many properties.

Chapter Three
The Properties of Manganese

Each element has a unique set of physical and chemical properties. These properties help scientists identify and classify individual elements. The physical properties of an element are those that can be measured or observed without combining the element with other substances. Some examples of physical properties are an element's color, its physical state at room temperature (whether it is a solid, liquid, or gas), its hardness, and its melting point. The chemical properties of an element describe the element's ability to undergo a chemical change or reaction with other substances. A chemical change alters the identity of the original matter.

Manganese's Appearance and Physical State

Manganese is a silvery white metallic element. It has no particular smell or taste. It looks and feels a lot like iron. At room temperature, all of the transition metals except copper (Cu) and gold (Au) are silvery in color. Manganese, like all of the transition metals, appears shiny when clean and polished.

The three physical states, or phases, of matter are solid, liquid, and gas. Manganese is found in the solid state at room temperature (68° Fahrenheit/20° Celsius). In fact, all of the transition metals except mercury (Hg) are solids at room temperature.

Transition metals may take many forms: a necklace made of silver (Ag), balls of nickel (Ni), a rod made of iron (Fe), a strip of copper (Cu), a gray lump of cobalt (Co), and a black lump of manganese (Mn).

Heating Manganese

If it is heated to a high enough temperature, solid manganese will turn into a liquid. Liquids are different from solids in that they do not have a fixed shape and they flow easily. The temperature at which manganese turns from a solid to a liquid is called its melting point. Manganese's melting point is 2,275°F (1,246°C). At an even higher temperature, it is impossible for the manganese to remain a liquid, and the liquid boils. This temperature is called the boiling point. Manganese's boiling point is 3,744°F (2,062°C).

Transition metals like manganese tend to have high melting and boiling points. This is usually due to the strong bonds that hold metal atoms

Manganese Snapshot

Chemical Symbol:	Mn
Classification:	Transition metal
Properties:	Hard, gray, brittle metal
Discovered by:	Carl Wilhelm Scheele, in 1774
Atomic Number:	25
Atomic Weight:	54.938 atomic mass units (amu)
Protons:	25
Electrons:	25
Neutrons:	30
State of Matter at 68° Fahrenheit (20° Celsius):	Solid
Melting Point:	2,275°F (1,246°C)
Boiling Point:	3,744°F (2,062°C)
Commonly Found:	In soil, meteorites, and on the ocean floor

together. Before a metal can change phases, these strong bonds must be broken. This takes a lot of energy, in the form of heat.

Manganese and Paramagnetism

Many of the transition metals are paramagnetic. A paramagnetic substance is one that is weakly attracted by a magnetic field. Manganese metal is paramagnetic due to the presence of unpaired electrons in its valence electron shells. Though manganese metal does not form a permanent magnet, it does exhibit strong magnetic properties when it is in the presence of an external magnetic field.

Manganese's Density

Density is another physical property that chemists can use to identify an element. Density measures the amount of matter contained in a specific space, or volume. The compound water (H_2O), for example, has a density of one gram per cubic centimeter (1.0 g/cm^3). Individual elements have unique densities. Transition metals generally tend to have high densities. Manganese has a density of 7.2 grams per cubic centimeter (g/cm^3). This tells us that a sample of manganese weighs 7.2 times more than a sample of water of the same volume. Because manganese has a greater density than water, shavings of manganese will sink if placed in water.

Chemical Reactivity and Manganese

If an element combines with other elements easily, it is said to be reactive. Manganese is a fairly reactive element. It actually behaves a lot like iron. Similar to iron, manganese slowly combines with oxygen in dry air. At higher temperatures, however, manganese reacts with oxygen more quickly. In some cases, it may even burn with the release of brilliantly

Underwater Treasure

In 1872, the British government sent out the HMS *Challenger* to investigate the oceans of the world. This began the gathering of some of our most important deep-sea knowledge. During its four years, the *Challenger* expedition discovered 4,417 new species of marine organisms. It also discovered manganese nodules lying scattered on the deep-sea floor, mainly in the Pacific Ocean. Nodules deep in the ocean have a manganese content as high as 36 percent. This is about the percentage of manganese that mining companies look for in the ore they mine.

Manganese nodules begin developing on things like a shark's tooth or a whale ear bone and grow extremely slowly. Scientists estimate that it takes several million years for the nodules to grow even half an inch (1.27 centimeters). These nodules also contain other metals like copper, cobalt (Co), nickel, and tungsten (W). The large number of nodules believed to exist on the ocean bottom makes the idea of deep-ocean mining attractive. Unfortunately, most of the high-quality nodules lie between 13,000 and 20,000 feet (3,962 and 6,096 meters) below the surface, making them very difficult to recover. The mining of these treasures was to begin between 1977 and 1980, but the plans were abandoned due to cost and politics.

These are manganese nodules deep on the bottom of the Pacific Ocean. Many of these nodules are as large as potatoes. It is estimated that manganese nodules cover from 20 to 50 percent of the Pacific Ocean bottom.

white light. In moist air, manganese corrodes, or rusts, perhaps even more readily than iron does. Manganese reacts slowly with cold water, but more readily with hot water or steam. Manganese also dissolves in most acids. The reactivity of manganese explains why it is never found as a free metal in nature—outside of meteorites—but is instead always found combined with other elements.

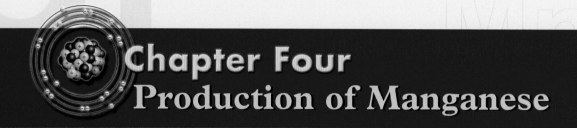
anganese is found throughout the world in the form of ores. Manganese ores are rocks that contain manganese metal and other elements. The most abundant ore of manganese is pyrolusite, or manganese dioxide. Pyrolusite is usually found in soft masses that stain the hands like charcoal. Other manganese ores include black manganite, black psilomelane, and rhodochrosite, which is pink. Most of the manganese mined today is found in South Africa and Ukraine. South Africa has about 80 percent of the world's minable manganese ore. Manganese mining has brought money and employment to areas of South Africa where cattle and sheep farming were the only other important sources of income. Other important manganese deposits are found in Brazil, China, Gabon, Australia, Russia, and India.

Making Steel Strong

Steel, which is made from iron, is used to make everything from skyscrapers to paperclips. In the 1850s, Henry Bessemer (1813–1898) devised the first inexpensive way to produce steel. His method refined, or purified, molten iron by blowing air through it. In the beginning, Bessemer ran into some trouble with his process. It was difficult for him to judge when the process was complete, so he sometimes continued to blow air into the

Pyrolusite, or manganese oxide, is a common ore of manganese. It usually appears as an earthy black mass. However, pyrolusite sometimes forms fernlike patterns (above), which are mistaken for fossilized plants.

steel, beyond what was necessary. As a result, the steel ended up with too much oxygen in it, making it hard and unworkable. But in 1856, Robert Mushet (1811–1891) solved Bessemer's problem when he added the alloy spiegeleisen to molten steel. (An alloy is a mixture of metals.) Manganese in the spiegeleisen combined with the unwanted oxygen, leaving the steel more workable. This solved Bessemer's problem, and from then on, Bessemer steel was widely accepted—thanks to manganese. In addition to removing oxygen, it was found that the manganese reduced the bad effects of sulfur (S) and phosphorus (P), two other impurities that can weaken steel.

Smelting Manganese Ore

Ferromanganese and other manganese alloys used for steelmaking are produced by a process called smelting. Smelting involves heating an ore to very high temperatures in order to obtain the metal. Ferromanganese, the main product of manganese smelting, is produced by heating the crushed manganese ore pyrolusite with iron oxide and coke (a type of coal that is almost pure carbon). Today, most manganese smelting is done in an electric furnace.

Much of the world's manganese is smelted for steel production. In addition to an electric furnace, another way to smelt manganese is to use a tall furnace called a blast furnace, shown here.

Manganese and Steel

Roughly 6,220,000 tons of manganese is produced each year. Between 80 and 90 percent of this manganese is used to make steel. Steelmaking would be impossible without manganese. In small amounts, manganese alloy is added to molten steel to remove the sulfur, phosphorus, and oxygen impurities from it. (Typically, about sixteen pounds of manganese metal are added to each ton of steel.) In large amounts, manganese is added to steel to make alloy steel. Steel is not considered alloy steel unless it contains a certain amount of other metals. Steel that contains 2 percent or more manganese is considered to be alloy steel. The first alloy steel was

Sneaking Up on the Enemy

Sometimes it is necessary for a submarine to sneak up on an enemy vessel or to evade one. Copper-manganese alloy with over 50 percent manganese vibrates a lot less than other metals. It is therefore used to make propellers for submarines. Less vibration makes for quieter subs that are not as likely to be detected by the enemy.

This picture, taken in 1954, shows one of four gigantic propellers belonging to the USS *Forrestal*, an aircraft carrier. The propellers were made of manganese bronze, a metal alloy that resists corrosion by seawater.

a high-manganese steel created in 1882 by Sir Robert Hadfield (1858–1940), a British metal expert. This steel, which contains 12.5 percent manganese, is now called Hadfield steel. Hadfield steel is very hard and resistant to impact. For this reason, it is used to make heavy-duty things, such as rifle barrels, prison bars, railroad tracks, safes, and equipment used for crushing.

Manganese and Other Alloys

Pure manganese is too brittle to be used by itself. However, when manganese is added to other metals, such as copper and aluminum (Al), it greatly increases the strength and corrosion resistance of those metals. Manganese is used in the preparation of many alloys. Of these, aluminum-manganese alloy ranks as one of the most important. The addition of manganese to aluminum enhances the metal's strength and resistance to corrosion. Aluminum-manganese alloys are used in kitchenware, roofing, and car parts. However, by far the most important use for aluminum-manganese alloys is beverage cans. Nearly 200 billion of these cans are produced worldwide each year!

Alloys of manganese and copper are used to make temperature-control devices in cars and other vehicles. Manganese bronze is made of manganese, copper, tin (Sn), and zinc. It is used to make propeller blades on boats and torpedoes because it resists corrosion by seawater. Another alloy, made up of copper, manganese, and nickel, is used to make small parts for watches. This alloy is sold commercially under the name Manifor.

Another important manganese alloy is silicomanganese, which is composed of silicon, manganese, and carbon. This alloy is a component in certain steels and is used to make structural parts and heavy-duty springs.

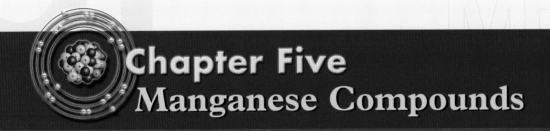

Chapter Five
Manganese Compounds

There are millions of different compounds all around you. A compound is formed when two or more elements bond together. The atoms of different elements bond with one another using their valence electrons. Some bonds involve the transfer of electrons between the atoms. These bonds are called ionic bonds. An example of an ionic bond is the bond that holds the manganese sulfate ($MnSO_4$) compound together. For this compound to form, the manganese (Mn) atom loses two electrons to the sulfate (SO_4) part of the compound. In other bonds, atoms share electrons between them. These are called covalent bonds. Manganese forms both ionic and covalent bonds. Since manganese is such a reactive metal, it forms a lot of different compounds. These compounds come in many striking colors. Although the most common use of manganese is steelmaking, manganese compounds have a variety of other uses.

Manganese Dioxide

Manganese dioxide (MnO_2), known as pyrolusite when found in nature, is the most plentiful of all the manganese compounds. Manganese dioxide is an important component in batteries. The Leclanché cell, often called a dry-cell or flashlight battery, was invented in 1866 by French chemist

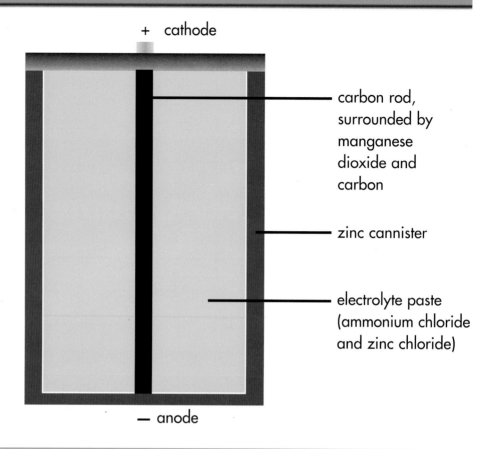

+ cathode

carbon rod, surrounded by manganese dioxide and carbon

zinc cannister

electrolyte paste (ammonium chloride and zinc chloride)

— anode

A simple diagram of a Leclanché cell. The manganese dioxide mixture surrounding the carbon rod is crucial to the chemical reactions that allow the cell to produce an electrical current.

Georges Leclanché (1839–1882). In a Leclanché cell, the negative electrode (called the anode) is made of zinc, and the positive electrode (called the cathode) is a carbon rod surrounded by a paste made of manganese dioxide and carbon. Manganese dioxide is also used in glassmaking to remove the green tint caused by iron impurities.

Manganese Sulfate

Manganese sulfate ($MnSO_4$) is a pink crystalline solid that is used for dyeing cotton. It is made by reacting sulfuric acid with manganese dioxide. Manganese sulfate is also included in multivitamins. Manganese

Toxic Manganese

Manganese compounds are less toxic than compounds of other metals, such as copper, iron, and nickel. However, in excess, manganese can cause problems. People who are exposed to high levels of manganese over a long period may develop manganism, a disease similar to Parkinson's syndrome. Symptoms include muscle tremors, shakiness, slurred speech, and drowsiness. Welders and others who work in factories that produce manganese metal are those most likely to suffer from manganism, from breathing in harmful manganese dust.

The term "manganese madness" was first used in the 1800s to describe the psychiatric condition of workers who had been exposed to high levels of manganese oxides. These workers displayed signs of schizophrenia (a psychotic disorder), hallucinations (seeing and hearing things that aren't really there), and compulsive behavior. Interestingly, a University of California study on prison inmates showed a higher concentration of manganese in the hair of violent offenders versus nonviolent ones.

Welders fuse pieces of metal together using heat. Welders working with manganese may suffer ill effects if they breathe in too much manganese dust.

is needed in the body for normal growth and health. Plants, too, need manganese to grow, so the element is added to commercial fertilizer in the form of manganese sulfate and another manganese compound called manganese oxide (MnO).

Potassium Permanganate

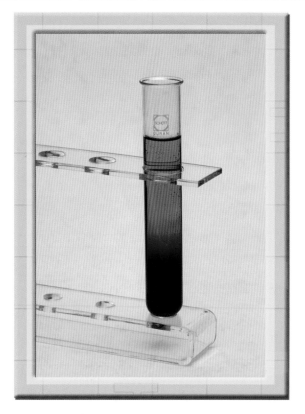

Above, the purple crystals of potassium permanganate are slowly being dissolved in a test tube of water.

Potassium permanganate ($KMnO_4$), also called Condy's crystals, is a dark purple crystal. In the right amounts, it can kill algae and bacteria without harming larger organisms, so it is often used to treat fish suffering with gill parasites or external fungal or bacterial infections. Potassium permanganate is also used to treat wastewater and toxic waste and to purify drinking water. It is used to control tastes and odors and to remove color. Potassium permanganate is also used for bleaching and removing color from fabrics. In concentrated form, potassium permanganate is even used to clear clogged drainpipes.

Manganese in Your Car's Gas Tank?

An important characteristic of gasoline is its octane rating. A gasoline's octane rating tells us how good the gasoline is at preventing knocking and pinging sounds in a car's engine. These sounds are caused by abnormal combustion, which occurs when fuel does not burn as cleanly or efficiently as it should. In 1958, lead was added to gasoline to reduce abnormal combustion. However, because it caused seriously harmful pollution, leaded gasoline was phased out in 1995. That same year, a manganese-containing compound became a legal fuel additive. Like lead, this compound, known as MMT, also helps reduce knocking and pinging in car engines. However, it appears that MMT may be a harmful pollutant, too. Some studies show that MMT is a powerful neurotoxin (a substance that damages nerve tissue) and a respiratory toxin.

Manganese Phosphate

Manganese phosphate ($MnPO_4$) is used in a process called parkerizing, or phosphating, to protect iron or steel surfaces from corrosion and to increase resistance to wear. Manganese phosphating involves dipping the metal part in a phosphoric acid solution whose key ingredient is manganese. The parkerizing process leaves a dark manganese-phosphate coating on the metal. This thick coating protects automotive parts, washers, fasteners, and other moving machine parts, as well as firearms.

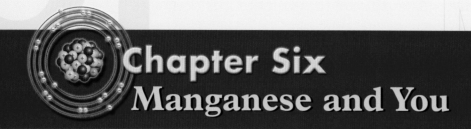

Chapter Six
Manganese and You

Manganese is a required nutrient for all plants and animals. The human body contains about 0.0007 ounces (20 milligrams) of manganese. Though this is a tiny amount, the element has very important roles.

Manganese in Enzymes

Manganese is found in many essential enzymes, which are biologically important compounds. Enzymes in your body speed up crucial chemical reactions. Without enzymes, many of the reactions that need to take place would be too slow to support life.

Enzymes containing manganese also protect the body from harmful oxygen radicals, highly reactive oxygen-containing molecules that can damage living cell tissues. Two examples of these protective enzymes are superoxide dismutase and pseudocatalase. (You can often recognize the names of enzymes because they usually end in "-ase.") Manganese-containing enzymes are also responsible for the body's proper use of carbohydrates, amino acids, and cholesterol. An enzyme called pyruvate carboxylase plays a critical role in the production of glucose, a sugar that our body uses for energy. Another manganese-containing enzyme, arginase, is needed by the liver to help break down and remove nitrogen from the body. This important process of nitrogen removal is called the urea cycle.

Let's Eat

Manganese is a component in both plant and animal tissue, so it is present in our food. Foods rich in dietary manganese include blueberries, olives, avocados, egg yolks, nuts, legumes, kelp (seaweed), tea, rice, green vegetables, and wheat. We need only a small amount of manganese to maintain good health. Infants and children require about 0.3 to 3 milligrams of manganese per day; adolescents and adults need about 2 to 5 milligrams. The typical person ingests an average of about 3.8 milligrams of manganese each day, though some take in much more than that. Too much manganese in the diet is generally not a problem, however. Most healthy people are able to excrete (get rid of) even moderately high excess amounts of consumed manganese. Manganese deficiency—not getting enough dietary manganese—does not normally affect humans, but in other animals it can cause problems in the proper growth of bone and cartilage. For this reason, manganese, in the form of manganese oxide, is often added to animal and poultry feed to promote good growth.

Manganese is called a trace element because it is normally found in very small quantities in rocks, soil, and water. (Other trace elements include calcium [Ca], iron, magnesium [Mg], cobalt, zinc, and copper.) The plants we eat get the manganese they need from the soil. However, some soil does not have enough manganese, resulting in stunted plants with yellow leaves. Manganese compounds, such as manganese sulfate and manganese nitrate, are added to fertilizers to help these plants grow. Manganese compounds are also sprayed on fruits and vegetables to kill harmful fungi that grow on them. One such fungicide is called mancozeb.

Plants need manganese to grow properly. The brown spots on these potato leaves are signs that this plant does not get enough manganese from the soil.

As mentioned, the cans in which a lot of our food and beverages are packaged are made of manganese-aluminum alloys and manganese-containing steel. We can also thank manganese for keeping our leftovers fresh. The aluminum foil that covers them contains manganese, too.

Planes, Trains, and Automobiles

Manganese is an important metal in the transportation industry, as it is used to make aircraft and vehicle frames. It is important for cars and trucks to be light because lighter vehicles use less fuel, and steel that contains manganese is stronger and lighter than pure steel.

Manganese is added to steel to strengthen it. In turn, steel reinforcement bar (or rebar) is used to strengthen concrete. The manganese-containing rebar shown here is being used on a large Chinese construction project.

Manganese is also used to make the roads and bridges on which we travel. It is found in the steel that makes rock crushers and other road construction equipment. For construction and industrial applications, manganese adds the important qualities of strength and resistance to wear. Manganese is also found in the steel that is used to make the structural beams that hold up highway overpasses and bridges. You can thank manganese for that the next time you use a bridge to get you safely across the water!

Manganese Around the House

Manganese is found throughout our homes, too. Aluminum-manganese alloys and alloy steel are very moldable, so these materials are shaped into our kitchen utensils or the siding on the outside of our homes or the awnings that decorate them. Manganese is also found in the electronic components of our television sets and audio systems. Manganese-steel alloys are found in appliances, such as dishwashers, and in our kitchen sinks, as well as in the steel pipes that bring our homes water and gas and remove sewage and waste. Even the foundations of our houses could contain manganese. To make sure that a foundation can withstand uneven forces from the soil, builders use reinforced concrete. The reinforcement comes from manganese-containing steel rebar, which is embedded in the concrete.

Let There Be Light

Safety matches can be ignited only when they are struck on a special rough surface on the side of the matchbox. They are safe because they do not self-ignite, and they are not poisonous. Manganese oxide in the head of these matches improves their performance.

The flashlight was invented in 1898, and it has since become a must-have gadget for both builders and homeowners. Flashlights would not be

Fireworks light up the sky with the help of manganese. Manganese dioxide (MnO_2) serves as an oxygen source for brighter lights. Manganese is also found in firework fuel.

possible without batteries and light-bulbs, and many batteries and lightbulbs would not be possible without manganese.

Fireworks also contain manganese. Fire needs oxygen to burn. It makes sense, then, that manganese dioxide would be used as an oxygen source for brighter lights. Manganese powder is also used in firework fuel to control or delay burn rates.

Manganese is an important element. It is vital in the making of steel, and we could not survive without its presence in our bodies. It is found in everything from batteries and electronics to house siding and cars. It is found in our television sets and in our lightbulbs. What a special element!

The Periodic Table of Elements

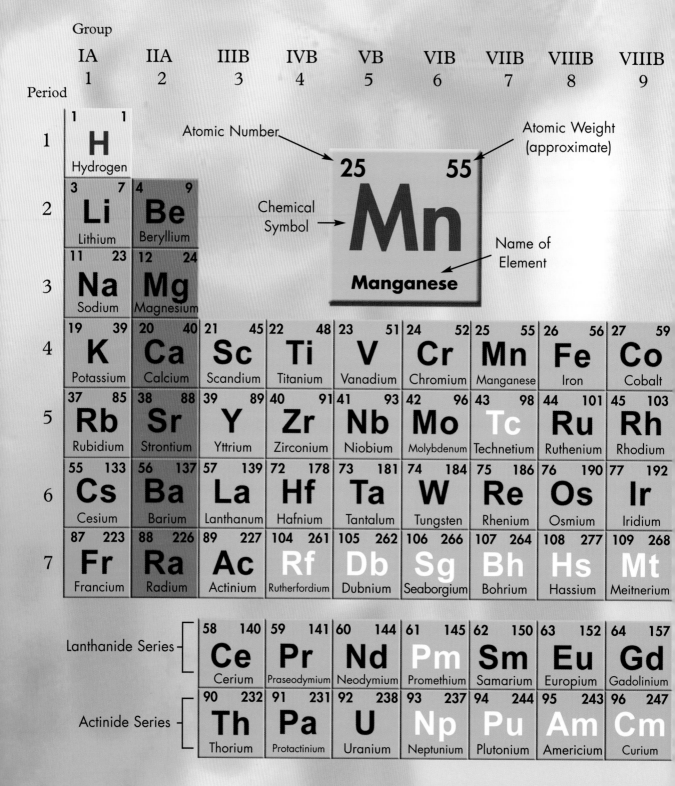

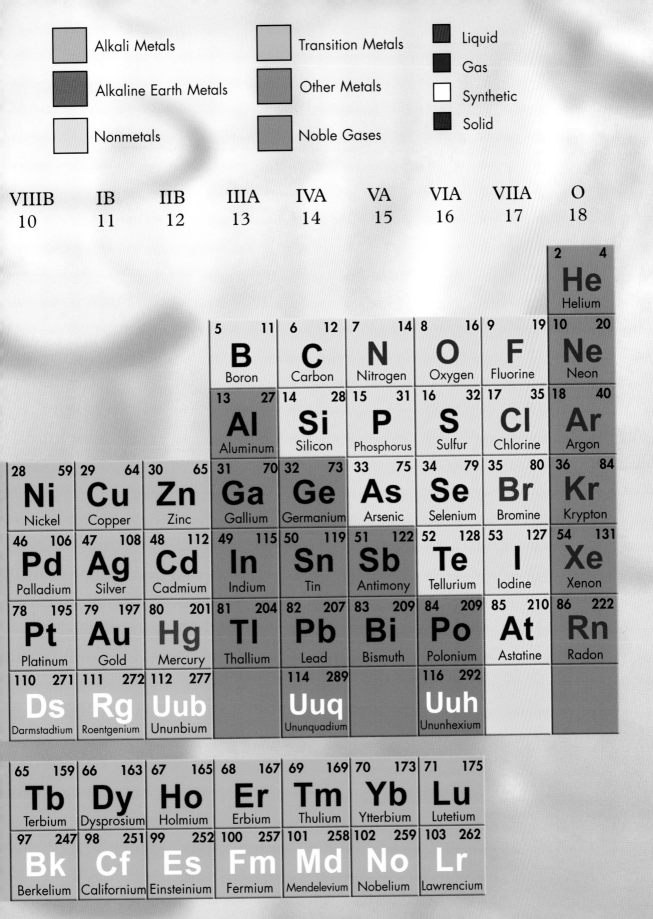

Legend

- Alkali Metals
- Alkaline Earth Metals
- Nonmetals
- Transition Metals
- Other Metals
- Noble Gases
- Liquid
- Gas
- Synthetic
- Solid

VIIIB 10	IB 11	IIB 12	IIIA 13	IVA 14	VA 15	VIA 16	VIIA 17	O 18
								2 **He** 4 Helium
			5 **B** 11 Boron	6 **C** 12 Carbon	7 **N** 14 Nitrogen	8 **O** 16 Oxygen	9 **F** 19 Fluorine	10 **Ne** 20 Neon
			13 **Al** 27 Aluminum	14 **Si** 28 Silicon	15 **P** 31 Phosphorus	16 **S** 32 Sulfur	17 **Cl** 35 Chlorine	18 **Ar** 40 Argon
28 **Ni** 59 Nickel	29 **Cu** 64 Copper	30 **Zn** 65 Zinc	31 **Ga** 70 Gallium	32 **Ge** 73 Germanium	33 **As** 75 Arsenic	34 **Se** 79 Selenium	35 **Br** 80 Bromine	36 **Kr** 84 Krypton
46 **Pd** 106 Palladium	47 **Ag** 108 Silver	48 **Cd** 112 Cadmium	49 **In** 115 Indium	50 **Sn** 119 Tin	51 **Sb** 122 Antimony	52 **Te** 128 Tellurium	53 **I** 127 Iodine	54 **Xe** 131 Xenon
78 **Pt** 195 Platinum	79 **Au** 197 Gold	80 **Hg** 201 Mercury	81 **Tl** 204 Thallium	82 **Pb** 207 Lead	83 **Bi** 209 Bismuth	84 **Po** 209 Polonium	85 **At** 210 Astatine	86 **Rn** 222 Radon
110 **Ds** 271 Darmstadtium	111 **Rg** 272 Roentgenium	112 **Uub** 277 Ununbium		114 **Uuq** 289 Ununquadium		116 **Uuh** 292 Ununhexium		

65 **Tb** 159 Terbium	66 **Dy** 163 Dysprosium	67 **Ho** 165 Holmium	68 **Er** 167 Erbium	69 **Tm** 169 Thulium	70 **Yb** 173 Ytterbium	71 **Lu** 175 Lutetium
97 **Bk** 247 Berkelium	98 **Cf** 251 Californium	99 **Es** 252 Einsteinium	100 **Fm** 257 Fermium	101 **Md** 258 Mendelevium	102 **No** 259 Nobelium	103 **Lr** 262 Lawrencium

Glossary

alloy Mixture of two or more metals.

bond Attractive force that links two atoms together.

chemical reaction Process that changes one kind of matter into another kind of matter.

compound Substance made up of two or more elements that are chemically bonded together.

element Substance made up of only one kind of atom.

enzyme Complex protein that speeds up biochemical reactions.

formidable Fearsome or awe-inspiring.

isolate To separate from another substance, so as to obtain a pure sample.

isotopes Atoms of the same element that have different numbers of neutrons.

manganism Debilitating disease caused by overexposure to manganese.

mineral Substance that forms naturally in rocks or in the ground.

nodule Small, irregularly shaped mass.

ore Mineral deposit from which metal can be extracted.

parkerize To cover a metal with a protective layer of manganese-phosphate.

periodic Occurring or recurring in a regular pattern.

phase Physical state of matter, either solid, liquid, or gas.

radical Atom or molecule with unpaired electrons.

smelting Process of melting a metal ore in order to extract the metal from it.

valence electrons Electrons in the outermost electron shell of an atom.

volume Amount of space that something occupies.

For More Information

Cave of the Mounds
2975 Cave of the Mounds Road
P.O. Box 148
Blue Mounds, WI 53517
(608) 437-3038
Web site: http://www.caveofthemounds.com

The International Manganese Institute
17 rue Duphot
75001 Paris
France
Web site: http://www.manganese.org

International Union of Pure and Applied Chemistry
IUPAC Secretariat
P.O. Box 13757
Research Triangle Park, NC 27709-3757
(919) 485-8700
Web site: http://www.iupac.org

San Diego Natural History Museum
1788 El Prado
San Diego, CA 92101
(619) 232-3821
Web site: http://www.sdnhm.org

Timpanogos Cave National Monument
Route 3, Box 200
American Fork, UT 84003
(801) 756-5238
Web site: http://www.utah.com/nationalsites/timp_cave.htm

Web Sites

Due to the changing nature of Internet links, Rosen Publishing has developed an online list of Web sites related to the subject of this book. This site is updated regularly. Please use this link to access the list:

http://www.rosenlinks.com/uept/mang

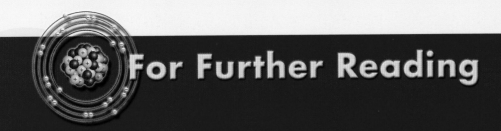

For Further Reading

Baldwin, Carol. *Metals*. Chicago, IL: Raintree, 2004.

Graham, Robin David. *Manganese in Soils and Plants*. New York, NY: Springer-Verlag, 1988.

Knapp, Brian. *Iron, Chromium and Manganese*. Danbury, CT: Grolier Educational, 2002.

Oxlade, Chris. *Metals*. Chicago, IL: Heinemann, 2002.

Saxton, Robert. *Manganese*. Manchester, England: Carcanet Press, 2003.

Stwertka, Albert. *A Guide to the Elements*. New York, NY: Oxford University Press, 2002.

Walshaw, Keith. *Iron, Chromium and Manganese*. Oxfordshire, England: Atlantic Europe Publishing Co. Ltd., 1996.

Bibliography

Brady, James E., and John R. Holum. *Chemistry: The Study of Matter and Its Changes.* New York, NY: John Wiley & Sons, Inc., 1993.

Calvert, J. B. "Chromium and Manganese." University of Denver. May 29, 2004. Retrieved June 29, 2007 (http://www.du.edu/~jcalvert/phys/chromang.htm).

Ebbing, Darrell D. *General Chemistry,* 4th ed. Boston, MA: Houghton Mifflin Company, 1993.

Heiserman, David L. *Exploring Chemical Elements and Their Compounds.* Blue Ridge Summit, PA: Tab Books, 1992.

Newton, David E. *Chemical Elements*, Vol 2. Farmington Hills, MI: UXL, 1999.

About the Author

Heather Hasan graduated from college summa cum laude with a dual major in biochemistry and chemistry. She has written books about other elements, including iron, nitrogen, aluminum, fluorine, and helium. She currently lives in Rockville, Maryland, with her husband, Omar; their sons, Samuel and Matthew; and their dog, Mofrey.

Photo Credits

Cover, pp. 1, 13, 15, 16, 40–41 by Tahara Anderson; p. 5 © Private Collection/© Look and Learn/The Bridgeman Art Library; p. 7 Erich Lessing/Art Resource, NY; p. 8 © Georg Gerster/Photo Researchers, Inc.; p. 10 © SSPL/The Image Works; p. 19 © Andrew Lambert Photography/ Photo Researchers, Inc.; p. 22 © Educational Images/Custom Medical Stock Photo; p. 25 © Lester V. Bergman/Corbis; p. 26 © AFP/Getty Images; p. 27 © Bettmann/Corbis; p. 31 © Getty Images; p. 32 © Martyn F. Chillmaid/Photo Researchers, Inc.; p. 36 © Nigel Cattlin/Visuals Unlimited; p. 37 © Sean Yong/Reuters/Corbis; p. 39 Shutterstock.com.

Designer: Tahara Anderson; **Editor:** Christopher Roberts